EVERY BIRD

IS A MIRACLE

❂

EVERY BIRD

IS A MIRACLE

✺

TEXT BY

TARA ROEDER

ILLUSTRATIONS BY

ARMAN SAFA

NEW MICHIGAN PRESS
DEPT OF ENGLISH, P. O. BOX 210067
UNIVERSITY OF ARIZONA
TUCSON, AZ 85721-0067

<http://newmichiganpress.com>

Orders and queries to <nmp@thediagram.com>.

Copyright © 2019 by Tara Roeder & Arman Safa.
All rights reserved.

ISBN 978-1-934832-67-7. FIRST PRINTING.

Printed in the United States of America.

Design by Ander Monson.

Cover and interior images by Arman Safa.

CONTENTS

Invitation To Join The Audubon Society   1
Figure 1: Whooping Crane   2
Thaw   3
Figure 2: Peregrine Falcon   4
Series In Which The Future Reveals Itself In Backwards Spurts   5
Figure 3: Finch   6
Los Angeles Is Burning   7
Figure 4: Hummingbird   8
Tiny Things   9
Figure 5: Tawny Owl   10
Tango Tuesdays   11
Figure 6: California Condor   12
Fig, The Ripened Ovary of Ficus Carica   13
Figure 7: Bald Eagle   14
Punching Mr. Tedesco   15
Figure 8: Pelican   16
Bower   17
Figure 9: Philippine Eagle   18
Around Valentine's Day   19
Figure 10: Wattled Crane   20
The Female Ornithologists' Club   21

Notes   24

INVITATION TO JOIN THE AUDUBON SOCIETY

I.

You find me in a dressing room wrapped in red-cheeked cordon bleus, their little wings whipping the air into liquid. You should, the woman watching us on the camera suggests tenderly, build them a collection of tiny beds on which to unwind.

II.

A friend of mine devoted themselves to ornithology at the tender age of 34. I imagined flocks of birds spreading like a parachute in the sky, thirstily catalogued.

Once I had a student who thought a starling was a baby star.

III.

We watch nature documentaries to remind ourselves that somewhere is a bird of paradise—two birds of paradise—inkily plumed, chests puffed out—desperately leaping in hope of a moment's copulation.

We watch nature documentaries to remind ourselves that somewhere is a swooping hawk about to make a kill, and we don't even know who to root for.

Figure 1: Orphaned Whooping Crane colt fed by Whooping Crane puppet on hand of disguised human

THAW

When you were young, lost in weeds, veined hand map alone to guide you, arms slowly becoming wings, iridescent marrow in your bones, plucking marbles for your nest—

When you returned in spring, beaked and plumaged, head tilted sideways, eyes black as coal, no one recognized your call, your tale of melt and flight.

Figure 2: Two orphaned Peregrine Falcon eyasses fed by Peregrine Falcon puppet on human hand

SERIES IN WHICH THE FUTURE REVEALS ITSELF IN BACKWARDS SPURTS

5.) In the year 3067, P. Smith's treasured collection of fossils and oddities includes a pickled human foot and seventeen hummingbird skeletons.

4.) In the year 3015, a sly creature emerges. The alley cats are awarded restitution.

3.) In the year 2090, a nouveau mystic is the first to report a miraculous and inexplicable occurrence involving an abandoned mechanical arm at the bottom of the sea.

2.) In the year 2061, Linus remembers shellfish. No one melts. The producer of the last great American sitcom is born.

1.) In the year 2022, two sundered ex-lovers do not recognize one another on a crowded London street. A tiny bird falls from the sky unnoticed.

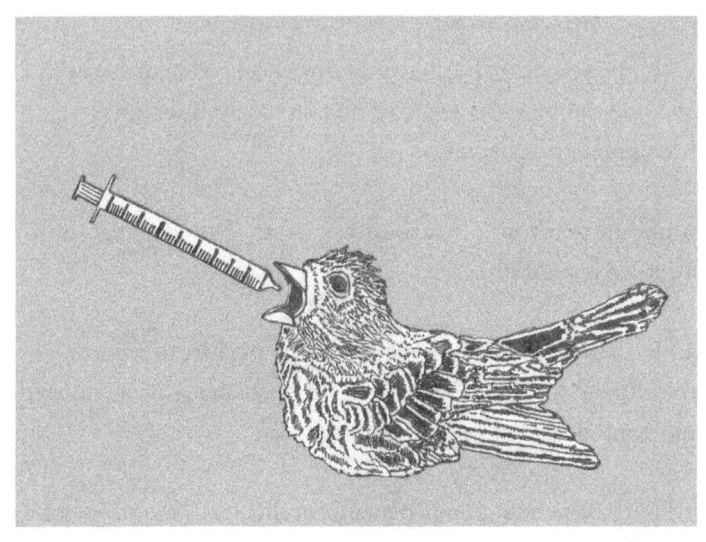

Figure 3: Orphaned Finch chick fed by dosing syringe

LOS ANGELES IS BURNING

Ashes fall like manna. The sooty ghosts of escaping angels choke the sky.

The two of you amass an army of extras from a zombie movie and make your way to the city limits. No one tries to stop you. The freeway is a river of molten ink.

"We've got this, babe," he says.

Your heart aches for the game shows that will never air, the uneaten shu mai, the remnants of the Hollywood sign.

A police helicopter drops from billowing red clouds like a stone and you tuck your hands in your pockets. Among the gum wrappers and ticket stubs you find an unsolicited fortune cookie.

Don't look back, you read out loud.

He can't help himself; he instinctively spins at the sulfurous explosion.

"No!" you cry, but the seal has been broken.

He becomes a pillar of salt. Golden finches halo what was once his head.

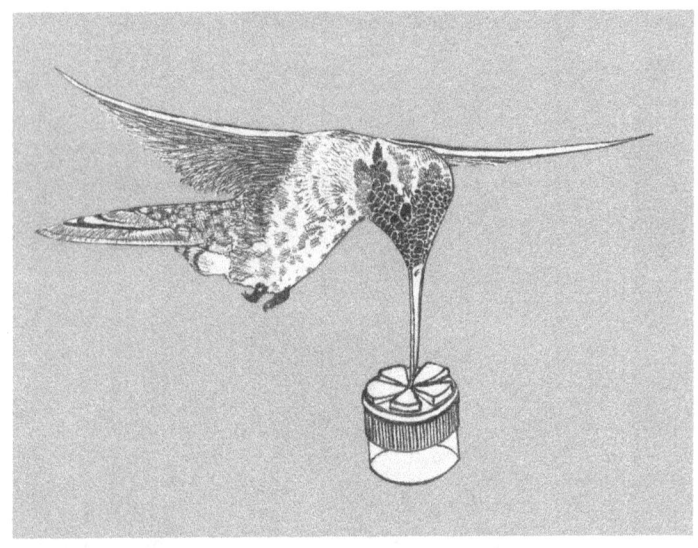

Figure 4: Orphaned Hummingbird chick fed by mini hummingbird feeder

TINY THINGS

In deep night tiny babies dream of tiny things—tiny shards of glass; tiny cactus blossoms; dust mites; diatomic hydrogen molecules. Last night they organized a miniature quilting bee, a charity event for a local doll hospital in desperate need of bedding. They heard a low buzz and turned to find a swarm of glistening hummingbirds eager to get started.

Figure 5: Pair of orphaned Tawny Owlets fed by Tawny Owl puppet on hand of human holding chopsticks

TANGO TUESDAYS

The couples dancing class was, in retrospect, the worst idea I've ever had. Worse than the rooster sanctuary. Tango Tuesdays—my alliterative downfall.

I didn't even know that there were astronauts anymore. Who could compete with her? Lunar boots and a rose in her mouth? That tiny, diaphanous dress? It just wasn't fair. I knew where you were the night you never came home. I didn't believe the story about the hospital, or the fake scar. I could picture the two of you floating together, bodies entwined, defying gravity.

When I emerged from the flooded basement that morning, I saw it in your eyes. A calculated, assessing look. The stars, or this waterlogged woman triumphantly gripping a monkey wrench as the one remaining rooster screeched his a.m. song?

No one could blame you. But if you answered my calls, you wouldn't regret it. When I said I was going to poison you, I didn't actually mean it and you know it. It was the gin talking.

Figure 6: Orphaned California Condor chick fed by California Condor puppet on human hand

FIG, THE RIPENED OVARY OF FICUS CARICA

Nobody listens to Cassandra. (Who won the space wars/the names of multiple medieval torture devices/why the clocks were melting/usurious percentages.) Floating fig trees drop offspring uneaten by neighborhood urchins who have been forewarned/spun into spiderwebs.

As a young girl, she was the recipient of a handsome collection of tightly bound instruction manuals sewn together with human hair: How to pickle figs. How to operate a table saw. How to find the perimeter of a large museum. How to ascertain the wingspan of the formidable pterodactyl, potential mother of modern birds.

Figure 7: Orphaned Bald Eaglet fed by Bald Eagle puppet on human hand

PUNCHING MR. TEDESCO

There's an episode of *Family Ties* in which several members of the Keaton family punch an incredibly rude English teacher in the face. I make my boyfriend watch this episode at one o'clock in the morning on a television channel that plays mostly 80s re-runs. We wonder if such an episode would fly now.

A few days later the Nazis are on the news. There are renewed debates about if it's ok to punch a Nazi.

They say hysteria is the product of a wandering uterus, but even after my hysterectomy I want to punch lots of people. Including but not limited to: a Nazi; my first gynecologist; people who keep birds in cages; every man who's ever complimented my stunning eyes.

I notice there are never any public debates about if it's ok to punch an English teacher. The default position is probably not.

Figure 8: Orphaned Pelican chick fed by human holding fish

BOWER

With ragged lightning breath a minor and forgotten god darts down from rolling clouds and kisses chosen lawns, which bloom incessantly.

Improbable flowers burst from the soil. Lopsided cakes of honey and whey. A microscopic wedding processional directed by two goldcrests.

He arranges a bower and invites you in. His fingers haven't forgotten. His primary materials are hydrangeas, broken crockery, and the discarded wings of molting pixies.

When the sun spills, you share a cigarette and swap stories. You were once a hotel maid, and he a jealous brother. Both dreaming of beaches coated in snow. Pineapples so ancient they explode at the touch. Apples the size of a fingernail.

He feeds you chervil stuffed tomatoes; quail eggs in aspic; ambrosia.

The sky is green; the moss grows quickly. In the morning he is gone, a wisp of salty smoke where he lay dreamless.

Seven years later, in the pocket of a forgotten vest, you find the seeds of an over-ripe pomegranate wrapped in a tendril of his hair.

Figure 9: Orphaned Philippine Eaglet fed by Philippine Eagle puppet on hand of human holding chopsticks

AROUND VALENTINE'S DAY

Trudy passes Vern and Linda's new house on her way to the diner. Her heart stops as she takes in the pink, plastic flamingoes, the wreath of artificial hyacinths adorning the door, the ornate birdbath inscribed with golden calligraphy: *V and L Forever.*

When she gets to work, she approaches Thelma, who is pouring decaffeinated coffee for two regular customers.

"You know those candy hearts?" Trudy asks. "*Kiss Me? Be Mine?*"

Thelma gives a tiny grunt of assent.

"If I were gonna print little messages on hearts they wouldn't be so sweet," Trudy says.

The women share a knowing glance.

But Trudy has a kind heart. "Not like *Screw You* or *Kiss My Ass*," she clarifies. "That would be too much."

Figure 10: Orphaned Wattled Crane colt fed by Wattled Crane puppet on hand of disguised human

THE FEMALE ORNITHOLOGISTS' CLUB

Welcome sisters.

We learn by doing here. We don't stumble into walls, our noses buried in Audubon guides. We interest ourselves solely in the alight and quickening. We secret ourselves in bushes. We lie in wait.

Make no mistake—every bird is a miracle.

That having been said, three North American species to watch out for:

1.) Red tailed hawk. Rumored to cause serious injury and mental anguish. One of their lot terrorized a Connecticut school in 2010.

2.) Mute swan. Territorial creatures, a gang once overturned a man's kayak and left him to drown.

3.) Great northern loon. One plunged his razor sharp beak into an ornithologist attempting to study him with deadly consequences.

But this is the minority of birds. Most are inquisitive, delightful creatures. Hummingbirds will entrance you. Bold types like seagulls will grab sandwiches right out of your hand.

A final word of advice: if you see a parrot, it is probably someone's missing pet. Proceed accordingly.

Now let's do this.

ACKNOWLEDGMENTS

In grateful acknowledgment to the venues in which these pieces (or versions of them) first appeared:

"Invitation To Join The Audubon Society" in *Bateau*; "Thaw" in *A-Minor Magazine*; "Series in Which The Future Reveals Itself In Backwards Spurts" in *Otoliths*; "Los Angeles is Burning" in *The A-3 Review*; "Fig, The Ripened Ovary of Ficus Carica" in *Clockwise Cat*; "Punching Mr. Tedesco" in *Permafrost Magazine*; "Bower" in *Leopardskin and Limes*; "Tango Tuesdays" in *Toasted Cheese*; "The Female Ornithologists' Club" in *Bateau*.

TARA ROEDER is an associate professor of writing studies in Queens, New York. She is the author of the dancing girl press chapbook *(all the things you're not)*, and her work has appeared in multiple venues including *3:AM Magazine*, *The Bombay Gin*, *Hobart*, and *Cheap Pop*.

ARMAN SAFA is a Brooklyn-based artist working primarily in ink, collage, printmaking, and assemblage. He manages an independent bookstore and is the founder of driftwood greetings, for which he designs greeting cards using collaged ephemera.

❁

COLOPHON

Text is set in a digital version of Jenson, designed by Robert Slimbach in 1996, and based on the work of punchcutter, printer, and publisher Nicolas Jenson. The titles here are in Futura.

❁

NEW MICHIGAN PRESS, based in Tucson, Arizona, prints poetry and prose chapbooks, especially work that transcends traditional genre. Together with DIAGRAM, NMP sponsors a yearly chapbook competition.

DIAGRAM, a journal of text, art, and schematic, is published bimonthly at THEDIAGRAM.COM. Periodic print anthologies are available from the New Michigan Press at NEWMICHIGANPRESS.COM.

www.ingramcontent.com/pod-product-compliance
Lightning Source LLC
Chambersburg PA
CBHW031508040426
42444CB00007B/1260